# ARMOR

BY ELIZABETH NOLL

BELLWETHER MEDIA • MINNEAPOLIS, MN

Torque brims with excitement
perfect for thrill-seekers of all kinds.
Discover daring survival skills, explore
uncharted worlds, and marvel at mighty
engines and extreme sports. In Torque books,
anything can happen. Are you ready?

This edition first published in 2022 by Bellwether Media, Inc.

No part of this publication may be reproduced in whole or in part without written
permission of the publisher. For information regarding permission, write to
Bellwether Media, Inc., Attention: Permissions Department,
6012 Blue Circle Drive, Minnetonka, MN 55343.

Library of Congress Cataloging-in-Publication Data

Names: Noll, Elizabeth, author.
Title: Armor / by Elizabeth Noll.
Description: Minneapolis, MN : Bellwether Media, 2022. | Series: Military
    science | Includes bibliographical references and index. | Audience:
    Ages 7-12 | Audience: Grades 4-6 | Summary: "Amazing photography
    accompanies engaging information about military armor. The combination
    of high-interest subject matter and light text is intended for students
    in grades 3 through 7"–Provided by publisher.
Identifiers: LCCN 2021051733 (print) | LCCN 2021051734 (ebook) |
    ISBN 9781644876282 (library binding) | ISBN 9781648346392 (ebook)
Subjects: LCSH: Body armor–Juvenile literature.
Classification: LCC U815 .N65 2022 (print) | LCC U815 (ebook) |
    DDC 355.82418–dc23/eng/20211022
LC record available at https://lccn.loc.gov/2021051733
LC ebook record available at https://lccn.loc.gov/2021051734

Editor: Betsy Rathburn      Designer: Jeffrey Kollock

Printed in the United States of America, North Mankato, MN.

# TABLE OF CONTENTS

# BULLETPROOF

The soldier must rescue the prisoners.
He jumps out of the truck and runs as fast
as he can. Guns blaze all around him.

The soldier wears armor over his fatigues. A bullet hits the armor. The force of the bullet knocks the soldier down. But he gets up again. The armor saved his life. Now he can rescue the prisoners!

# WHAT IS ARMOR?

Armor is a special covering that protects the body from attacks. The military uses it to protect troops. Troops wear armor with their uniforms.

Armor can be made of many different materials. Early armor was made of metal pieces or thick leather. Today, many troops wear armor made of superstrong plastic.

# TIMELINE

## 4TH CENTURY BCE

## 11TH CENTURY BCE

LEATHER ARMOR

CHAIN MAIL ARMOR

1300s CE

1970s

METAL PLATE ARMOR

KEVLAR ARMOR

Different kinds of armor stop different kinds of weapons. Some armor stops knives or other sharp objects. It is usually a vest made to cover the upper body.

VEST

This armor is usually made of strong plastic.
Tiny plastic pieces overlap like scales.
They can stop sharp blades!

ARMORED VEHICLE

### ARMORED VEHICLES

The military also uses armored vehicles! They are made with strong metal, glass, and other materials.

Troops may also use armor during riots. Riot armor is made to protect against glass, rocks, and other heavy objects.

Riot armor often includes a heavy vest with many pockets. These hold bulletproof panels. Troops wear strong helmets to protect against thrown items. They may also carry long shields. These help control the movements of crowds.

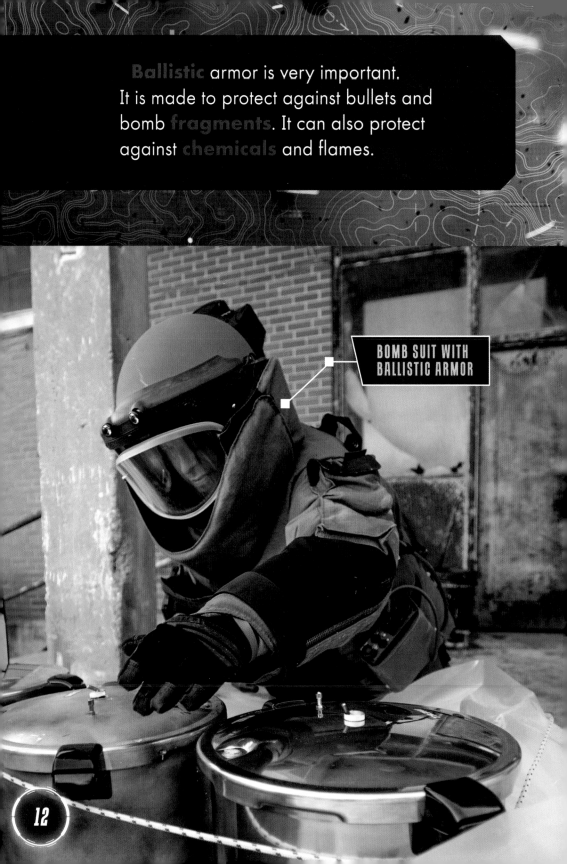

Ballistic armor is very important. It is made to protect against bullets and bomb fragments. It can also protect against chemicals and flames.

BOMB SUIT WITH BALLISTIC ARMOR

## MODULAR SCALABLE VEST

**RELEASED:** FIRST USED IN 2018

**FEATURES:** ARMOR PLATES TO PROTECT FROM BULLETS

Most ballistic armor is made from strong plastic. The plastic is used to make vests, helmets, gloves, and other gear.

# THE SCIENCE BEHIND ARMOR

KEVLAR

Scientists study materials to make the best armor. Kevlar is one of the most common. It is made of strong plastic fibers woven into a stiff, dense net. It is stronger than steel!

# HOW KEVLAR WORKS

**1**

KEVLAR

BULLET

**2**

**3**

= ENERGY DISPERSING

**INVENTED BY...**

Stephanie Kwolek invented Kevlar. She was trying to find a way to make car tires stronger. Now her work is used to make stronger armor!

When bullets hit Kevlar, the woven plastic helps disperse their energy. The energy spreads out. Layers of Kevlar keep bullets from reaching a soldier's body!

SAPI POCKET

SOLDIER PUTTING ON
KEVLAR VEST

Kevlar works with other materials to add
more protection. Kevlar vests are soft armor.
They feel like fabric and have many pockets.
The pockets hold SAPIs.

SAPIs are made of strong ceramic. This material resists bullets. If it is damaged, it can be replaced easily. Kevlar and SAPIs work together to make safer armor!

SAPI

# THE FUTURE OF ARMOR

GRAPHENE ARMOR

Scientists are working on future armor.
One possibility is liquid armor. It is lighter
and stronger than other armor. Kevlar is
soaked in a special liquid. If a bullet hits
the liquid, the liquid hardens instantly!

# FUTURE
# ARMOR PROFILE

## GRAPHENE ARMOR

**DEVELOPED:** BEGAN IN THE EARLY **2000s**

**FEATURES:** STRONG, LIGHT ARMOR THAT BECOMES AS HARD AS A DIAMOND UNDER PRESSURE

Future armor could also be made with graphene. It is strong, light, and flexible. Troops wearing this armor can move more easily!

Scientists are also working on spider silk armor. This lightweight material comes from spiders. It can be woven into a strong fabric that stops bullets!

# NON-MILITARY USES

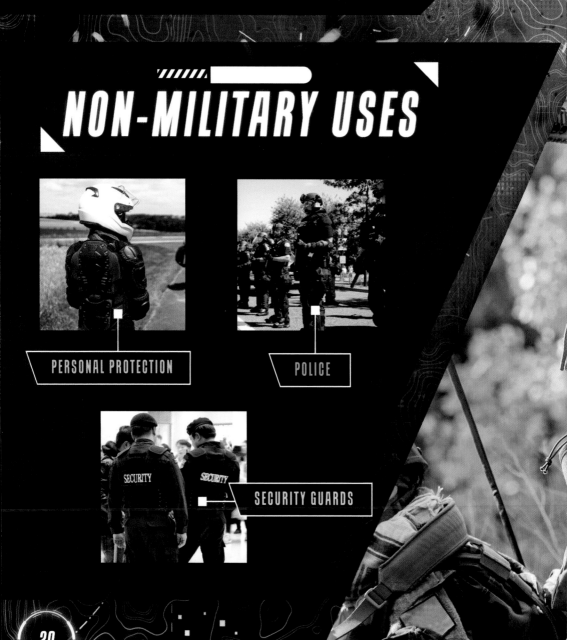

PERSONAL PROTECTION

POLICE

SECURITY GUARDS

Troops depend on armor to keep them safe.
As weapons get more powerful, armor must get
stronger. The future of armor is full of possibilities!

# GLOSSARY

**ballistic**—related to bullets

**ceramic**—a material made from clay

**chemicals**—substances created when two or more substances interact with one another

**dense**—made from materials that are very close together

**disperse**—to scatter or move away from the center

**fatigues**—the clothing a soldier wears

**fibers**—threads that make up a material

**fragments**—small pieces

**graphene**—a thin material made from carbon; graphene may be used to make strong, lightweight armor.

**Kevlar**—a bullet-resistant material woven from super hard plastic

**panels**—flat parts

**riots**—events caused by a crowd that disturb the peace

**SAPIs**—small arms protective inserts; SAPIs are ceramic plates that go inside an armor vest.

# TO LEARN MORE

## AT THE LIBRARY

Chandler, Matt. *Advanced Weaponry*. Minneapolis, Minn.: Bellwether Media, 2022.

Hofer, Charles C. *Beastly Armor: Military Defenses Inspired by Animals*. North Mankato, Minn.: Capstone Press, 2020.

Vonder Brink, Tracy. *The United States Army*. North Mankato, Minn.: Pebble, 2021.

## ON THE WEB

# FACTSURFER

Factsurfer.com gives you a safe, fun way to find more information.

1. Go to www.factsurfer.com

2. Enter "armor" into the search box and click 🔍.

3. Select your book cover to see a list of related content.

# INDEX

The images in this book are reproduced through the courtesy of: Getmilitaryphotos, front cover;
Nesterenko Maxym, p. 3; John Yountz/ DVIDS, pp. 4-5, 5; Huangdan2060/ Wikipedia Commons, p. 6
(leather armor); Miriam Doerr Martin Frommherz, p. 6 (chainmail); Anastasiya Aleksandrenko, p. 7
(plate armor); SoldiMaks, p. 7 (Kevlar); FXQuadro, pp. 6-7; PRESSLAB, pp. 6 (inset), 20-21; Trzykropy,
p. 8; DVIDS, pp. 8-9 (vehicle), 14-15, 16-17 (left); Quinton Batchelor, pp. 10-11 (left); Frederic Legrand
- COMEO, pp. 10-11; Jacob Bender/ DVIDS, pp. 12-13; U.S. Army/ Wikipedia Commons, p. 13 (vest);
Fotokostic, p. 13 (plates); KG Design, p. 14 (Kevlar); AB Forces News Collection/ Alamy, p. 16 (inset);
Puttawat Santiyothin, pp. 16-17 (right); Dorxela, p. 18 (inset); Digital Storm, pp. 18-19; Forance, p. 19;
Sergey Nemirovsky, p. 20 (personal protection); Eric Crudup, p. 20 (police); robert paul van beets, p. 20
(security); Dja65, p. 22.